Maya Math Simplified

by Njord Kane

Maya Math Simplified

By Njord Kane

© 2013, 2015 by Njord Kane. All rights reserved.

No part of this book may be reproduced in any written, electronic, recording, or photocopying form without written permission of the author, Njord Kane, or the publisher, Spangenhelm Publishing. You must not circulate this book in any format.

Books may be purchased by contacting the publisher and author at: spangenhelm.com

Published on: March 1, 2015 by Spangenhelm Publishing

Interior Design and Cover by: Njord Kane

Library of Congress Control Number: 2015920144

ISBN-13: 978-1-943066-087

ISBN-10: 1-943066-08-6

1.Maya 2.Mayan 3.Mathematics 4.Math 5.Ancient Civilizations

Second Edition.

10 9 8 7 6 5 4 3 2

 Spangenhelm Publishing
United States

Table of Contents

Preface..................................i

Chapter 1
Who were the Maya?......................1

Chapter 2
Ancient Maya Arithmetic................17

Chapter 3
The Value of Zero......................29

Chapter 4
The Four Slave Example.................37

Chapter 5
The Grid System........................43

Chapter 6
Subtraction............................57

Chapter 7
The Finger Method......................63

Chapter 8
The Maya Abacus........................75

Chapter 9
Maya Concept of Fractions 93

References............................101

Preface

The Maya were a major indigenous pre-Columbian civilization of the Yucatan Peninsula and are members of a modern American Indian people of southern Mexico, Guatemala, and parts of Honduras who are the descendants of this ancient civilization.[1]

Which is correct to use when referring to these people, is it 'Maya' or is it 'Mayans?' Is it a 'Maya' or a 'Mayan' archaeological site? We see the words, Maya and Mayan used interchangeably without discrimination. So, which is correct, do we use Maya or Mayas or Mayan or Mayans?

The adjective 'Mayan' is used in reference to the language or languages, whereas the noun "Maya"[**mah**-*yuh*][1] is used when referring to the people, places, and or culture, etc., without distinction between singular or plural. This convention is the most widespread among Mayanists (scholars who study and write about the Maya). This distinction arose in the field of linguistics, where the "Mayan" adjective started to be used to define the linguistic family that incorporates the different dialects spoken by the Maya people. In sum, "Mayan" are their languages and "Maya" for everything else in reference.

The Maya, like other ancient civilizations, used mathematics everyday in their day to day activities as well as improved math and even geometry to build great works, such as their pyramids and temples. They used mathematics for their calendars and in astronomy when they recorded the movements in the skies.

The Maya temples and pyramids still stand today and their sky charts are still accurate to this day. Something modern man has not been able to achieve until the use of computers were available to assist. That leaves a lot to say about the mathematics of these ancient people. To create celestial charts such as did the Maya people requires centuries of dedication. Learning and recording, passing on the knowledge and then adding upon it.

Building great cities with plazas, city centers, and great temples to stand against the elements and time required a firm understanding in geometry. The Maya had planned and constructed their buildings that not only used geometry, but also had their angles aligned with celestial events.

Chapter 1

Who were the Maya?

The Maya are an indigenous people whose culture had built a thriving ancient city-state civilization in Mesoamerica.

MesoAmerica is the location that lies in the area from Mexico to South America. An area considered to be the 'middle' of the Americas and is also known as the Central Americas.

Along with the Maya, there are many other indigenous cultures in the Mesoamerican area. Some of these other cultures are the Mexica (Aztecs), Mixtec, Purepecha, Huastec, Olmac, Toltec, Zapotec, and Teotihuacan.

These indigenous Mesoamerican cultures are credited with the creation and innovation of many inventions. They used advanced mathematics to engineer and build great pyramid temples that still stand after thousands of years. They were clear masters of observed astronomy and created highly accurate calendars.

They maintained stable enough societies to allow the practicing of fine arts and integrated it into a complicated writing system that balanced both math and writing into a complex theology. The Maya are credited as being the first culture in the New World to utilize a fully developed written language.

They practiced elective medicine and for the most part, used an intensive agriculture system to maintain huge populations.

The Mesoamericans had discovered the wheel, but the absence of draft animals and an often demanding terrain made human labor the most utilized means for the transportation of goods and building materials. Suitable bovine or equine were not introduced into the Americas until later when Europeans brought them over.

The areas dominated by the Maya are known today as the southern Mexican states: Chiapas, Campeche, Yucatan, Quintana Roo, and Tabasco. The Maya civilization spread all the way through the nations of Guatemala, Belize, El Salvador, and Honduras. A very large expanse of city-states that ruled the area linked by trade routes.

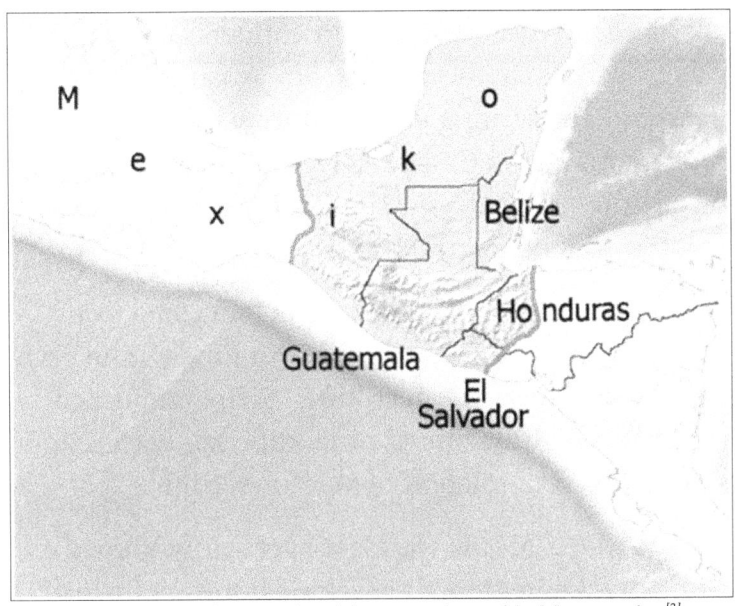

Map showing where Ancient Maya were located in Mesoamerica.[2]

Descendants of the ancient Maya civilization live today in the Yucatán Peninsula of Southern Mexico, Guatemala, and parts of Honduras and El Salvador.

The proximity of the Mesoamerican people to each other in the region led to a high degree of cultural interaction between each other. The consistent interaction between Mesoamerican civilizations within the region created a cultural diffusion that allowed Mesoamericans to share a great degree of their cultural practices and knowledge with each other. Knowledge in mathematics was also exchanged between these cultures.

Mesoamericans continually influenced each other, even when their interaction wasn't always peaceful. The writing and epigraphy used to create the famous 'Maya Calender' weren't even of Maya origination. They had assimilated it into their own culture from neighboring cultures in their region.

The writing used in the region had come from previous cultures and evolved over time within each different Mesoamerican culture. Script and usage becoming slightly altered or modified as each unique scribe used it in relation to their own culture.

The Maya people were not necessarily known as being great inventors themselves, but were instead great innovators that absorbed others advancements and continued to develop upon them within their own culture. The culture of the ancient Maya seemed to promote the application of inventions of the many other nearby cultures in the area and sought ways to improve upon them on their own.

Like many of the other Mesoamerican cultures, the Maya did not have a separation between religion and government. Church and State were one of the same. They considered the gods to be the everyday rulers of their daily lives and depended on their priests and rulers to ensure that the gods were appeased and didn't destroy the earth or extinguish the essential life sustaining Sun.

The Maya religion required a highly complicated method of worship that demanded bloodletting and sacrificial rituals that were often fulfilled by the kings and queens. These efforts were necessary because it was believed to "feed" the gods. It was the sacred duty and responsibility of the ruler to often feed the gods with their own blood. The believed their rulers had the power to pass in and out body to the spirit world and acted as messengers to the celestial world.[3]

Geographically, the Maya were formed individually as independent city-states. They used a government structure that allowed their individual rulers a great deal of individual governance within their own municipalities, instead of a strong centralized governing structure ruled by an emperor or empress.

The Maya civilization wasn't a single unified empire, but were instead a multitude of separate entities that shared a common cultural background. They shared several similarities with the Greeks, in that the Maya were religiously and culturally a nation, but were politically separate sovereign city-states.

The Tikal city center, one of the most powerful Classic Period Maya cities.[4]

Maya city centers were the epicenters for trade, religious, and other cultural activities which also included some local administration.[5] There were many Maya cultural centers located in what's considered "the Maya Area" that spreads across a large expanse covering a wide range of climate conditions.

Their culture spanned across mountain ranges into semi-arid plains and reached into the thick labyrinths of the rain forests. A diverse area that allowed for a diversity of trade.

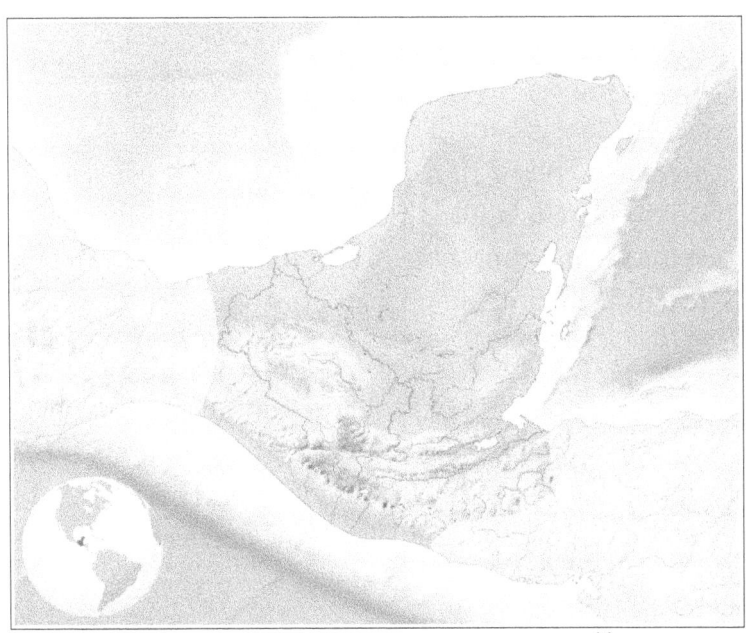

Map of the Maya Area in the Yucatán peninsula.[6]

The period of time before the arrival of Christopher Columbus and European expansion to the Americas is called the 'Pre-Columbian Period.' The Pre-Columbian period of Maya history divides into five distinct time periods.

- The Paleo-Indian Period ("First People" - 3500 BC),
- The Archaic Period (3500 BC - 2000 BC),
- The Preclassic Period (2000 BC - 200 AD),
- The Classic Period (200 AD - 900 AD),
- The Post Classic Period (900 AD - 1697 AD).

It was during the Paleo-Indian period when early nomads crossed into the Americas over 15,000 years ago. These were the "First People" to inhabit the Americas. They'd first crossed into North America until eventually splitting off from other groups and eventually migrating south through Mexico into the Yucatán Peninsula of Mesoamerica.

These migrating "First People" in the Maya region developed their tool and hunting technologies and went from being nomadic hunter-gatherers into forming more permanent settlements. These settled groups became more developed as they exploited the plentiful local resources.

These now settled groups progressed into the Archaic period and began advancing into a more complex society. These archaic settlements developed culture and technology that was shared with neighboring settled groups. The exchange of ideas between these groups formed into a shared culture that began developing into a culturally distinct people.

The Maya Civilization originated in the Yucatán region during the Preclassic Period at around 2000 BC. There is some argument as to when the Preclassic Period began for the Maya. It's argued to have began as late as 2600 BC, while there's claim that it's earlier because there are permanent Maya settlements along the Pacific coast that date to 1800 BC. A difference of eight hundred years, depending on region.

The Preclassic period begins where the first signs that the Maya can be recognized as a distinct people. The two time periods overlap each other as a result from different groups in the region gradually shifting from being a separate archaically developed people into adopting local culture and technology that was distinctly Maya.

It was also during the Preclassic period that the Maya developed a greater interest in art and began some degree of manufacturing. A number of Preclassic Maya pottery and clay figures that were fired in primitive kilns survive to this day. Many of these clay and pottery artifacts, that are well over four thousand years old, give us clues as to their origin and purpose. Indicators as to how advanced their technology was growing. The process of using buildings as a means of recording history had also began to develop during the Maya Preclassic era.

A very distinct Maya culture with religious beliefs and practices, as well as shared technologies, began to rapidly form and progress during the Preclassic period. Public ceremonies and rites begin taking place during the Preclassic period. The creation of burial rites for the dead began during this time. The Maya civilization continued to grow and advance into its Classical Period, where it reached its peak in development at around 200 - 250 AD. Still almost two thousand years before contact with Europeans.

The Classic Period for the Maya culture occurred from 200 to 900 AD. During this time, the Maya began to develop urban centers that were more focused on the pursuit of artistic and intellectual development. These city centers became true cultural hubs in various Maya city-states. Written documents from the Classical period demonstrate a highly developed method of communication amongst the Mesoamerican people. It was during the Classical period that engineering feats, such as the construction of pyramids in the city-states began emerging.

The desire to preserve personal and cultural histories begins to develop during the Classical period as well. There are many carved slabs of stone known as 'stelae' that have survived to tell the stories and lineage of important rulers of the time.

The Maya had developed a complex system of carved hieroglyphs to preserve the stories of historical events.

Maya Stelae and Pyramid located at the Copan Ruins in Honduras.[7]

Lidded effigy container in the form of a god, Late-Postclassic period.[8]

Towards the end of the Classic Period was when the structure of Maya society began to change. Settlements in the southern lowlands started dwindling in population until eventually becoming abandoned. This may be perhaps to natural disasters such as hurricanes known to the region.

The architecture began appearing seemingly ordinary rather than having the elaborately ornate inscriptions that were apart of the buildings built centuries prior. Building took on a more utilitarian emphasis rather than the previous . There were few, if any, grand structures appearing in the 8th or 9th centuries leading into the Maya Post-classic Period.

During the Post-classic Period, the Maya people and their culture continued to thriving in the Northern sections of the Yucatan' area. Buildings in new settlements were now being constructed with plain straight walls and flat ceilings. These simple lines now characterized the construction of new buildings, in contrast to the elaborate carvings and decorations used in construction during the previous period.

The earlier interest in art continued to be part of Maya culture as well as a continued interest in language and writing, Yet the great bursts of creativity that came out during the earlier periods appear to have ceased during the Post-classic period of the Maya civilization.

Assimilation with other neighboring cultures had weakened some of the distinctiveness of Maya culture as they interacted more heavily on neighboring cultures. Nevertheless there were still several city-states that retained a decidedly distinctive Maya culture well into the 16th century.

During the Post-classic period the Maya civilization continued as a major dominating force for 700 more years until around 900 AD when their culture became less dominate in the region.

The Maya city-states continued through the arrival of the Spaniards in 1511 AD and continued until after almost two centuries of efforts by Spanish Conquistadors, the last Maya city was conquered in 1697 AD.

Even after the Spanish Conquest and subsequent colonization, the Maya people and the spoken Mayan language did not die out with the end of their civilization. The legacy of the Maya civilization lives on today in several ways. Many members of the rural populations in Chiapas, Guatemala, Belize, and the Yucatan Peninsula are Maya by descent and utilize one of the Mayan dialects as their primary verbal language.

The Culture of the Maya people can be found influencing many cultures around Mexico and other parts of Central America. Artifacts that are undeniably of Maya origin have been found as far away as Central Mexico, which is more than 1000 kilometers away.

Chapter 2

Ancient Maya Arithmetic

The ancient Maya used a mathematical system that is "vigesimal." A vigesimal counting system is based on 20 units (0 - 19), instead of the 10 unit (0 - 9) based counting system that we use today called the decimal system.

The decimal mathematical system widely used today is believed to have possibly originated by counting the number of fingers that the average person has. Counting with our fingers gives us our ten unit based metric system. It is believed that the Maya possibly began counting with both their fingers and toes, which gave them their twenty based 'vigesimal' system. Their counting system is based on groups of twenty, instead of our modern ten.

When we count using our decimal system, we count to ten and then we add one value to the next tier, or level with a value of ten (10, 20, 30, etc.) for each cycle we reach on tier one (0-9). Using the decimal system, the count of 11 is the value of '1 unit' from "tier 2" and '1 unit' from "tier 1," making the total value equal to 11 (10 + 1 = 11).

Each tier in the decimal system counts up in multiples of ten. 1 x 10 = 100 x 10 = 1,000 x 10 = 10,000 and so on.

tier 3 =	100,	200,	300,	400,	500,	600,	700,	800, 900
tier 2 =	10,	20,	30,	40,	50,	60,	70,	80, 90
tier 1 =	1,	2,	3,	4,	5,	6,	7,	8, 9

Decimal tiers. 1 value from tier 2 is 10 and 1 value from tier 1 is 1, One value each from tier 1 and 2 equals 11 (10 + 1 = 11).

Although we users of the modern decimal system are not used to thinking of our counting in tiers or levels, it does help us understand how the count using the Maya vigesimal system. Just as the decimal system goes by tiers of ten: 1, 10, 100, 1000, 10000, etc., the Maya vigesimal system goes by tiers of twenty: 1, 20, 400, 8000, 160000, etc..

tier	decimal	vigesimal
6 -	100,000	3,200,000
5 -	10,000	160,000
4 -	1,000	8,000
3 -	100	400
2 -	10	20
1 -	1	1

Comparison of decimal and vigesimal tiers.

The modern decimal system has ten possible digits for each placeholder in a tier, numbering from 0 to 9. In the Maya vigesimal system, the first tier's placeholders has twenty possible digits, numbering from 0 to 19.

decimal tier 1
0 1 2 3 4 5 6 7 8 9

vigesimal tier 1
0 1 2 3 4 5 6 7 8 9 10 11 12 13 14 15 16 17 18 19

Linear comparison of first decimal and vigesimal tiers.

When the last number of a tier is reached, the count in each system proceeds up to the next tier. Adding a value of 10 in the decimal system or 20 in Maya vigesimal system.

decimal tier 2
10 20 30 40 50 90
vigesimal tier 2
20 40 60 80 100 380

Linear comparison of second decimal and vigesimal tiers.

For the number "21," both systems used the twenty from tier 2 and the one from tier 1 to make: 20 + 1 = 21. Except with the Maya vigesimal system, they used 1 value on tier 1 and 1 value on tier 2. The decimal system uses 1 value on tier 1 and 2 values on tier2.

To make "11," the decimal system uses 1 value from tier 2 equaling 10 and 1 value from tier 1. (10 + 1 = 11). The difference in the Maya vigesimal system that value of 11 is actually calculated with two 'fives' from tier 1 and an 'one' also from tier 1. On their system, if the count hasn't reached or passed the value of 20, it remains on the first tier. Additionally, the value of 20 was divided by 4 values of 5.

To represent these values, the Maya came up with two symbols of which they use, a dot and a bar. The Maya used a system of bars and dots for their numbers, instead of representing them by different symbols as it is done in our numbering system. A system the we adapted from Latin and Arabic. Each 'dot' in the Maya system represents the value of 1 unit on a given tier. Each bar represents the value of 5 dots of the tier below it.

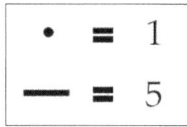

1 dot equals 1 value and 1 bar equals 5 dots.

So when counting to the value of 11, the Maya counted two bars and one dot from tier one (5 + 5 + 1 = 11).

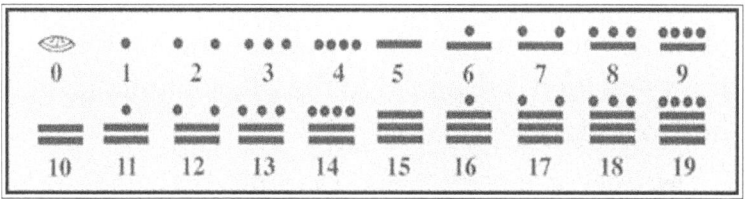

Two bars and one dot on tier 1 equals eleven. The dots go about the bars,

The Maya used a either a cocoa seed pod or a shell to represent the value of 0. They used a system of stacked bars and dots to represent the first 20 numbers.

Table showing first 20 Maya numbers and their Arabic equivalents.[9]

It is believed that because the cacao bean was commonly used as currency throughout

Mesoamerica, the Maya symbolized it with a dot to represent one bean.

Another reason to believe that the Maya got the idea of using a dot from the cacao bean, is the fact that they packaged their cacao beans in quantities of 8000 to a bag.[10]

The sum of 8000 is one of the place values on the Maya vigesimal system (20^3). The Maya name for the value of 8000 is called a "pic." This is also the name for the sack they used to pack cacao beans in. It's understanding why the Maya found it convenient to call value of 8000 a "pic" and the sack they used that head 8000 cacao beans. A set standard among the Maya for an amount in trade.

The Maya represented written values in a vertical manner, whereas our modern representations of number values are expressed horizontally. For example, we write the number 27 horizontally with the number two then the number seven to the right of it. As the numeric value increases, the number representations are added to the left, continuing horizontally.

The Maya, however, would write the value representation of 27 vertically. Their symbol for seven (a line representing five units with two dots over it) would be on the bottom and the symbol for 20 (a dot on the line above) would be directly over it.

The same applies for other numbers. The Maya script or sculptor would fashion their own style of glyph, but the marked value would still be written from bottom to top.

0	1	2	3	4
5	6	7	8	9
10	11	12	13	14
15	16	17	18	19
20	21	22	23	24
25	26	27	28	29

Maya horizontal positional value representations.

Maya positional counting displayed values of 20 or more by placing a value symbol over another value symbol. When writing with the Maya vertical vigesimal positioning system, the value of 20 is written with a shell representing zero placed at the base on the bottom position and a single dot, which represents the value of twenty is placed over it in the second vertical position.

A single dot in this position over a zero means one unit of the second tier equaling 20. To write 21, the zero on the first tier would be changed to a single dot (1 unit) and for the subsequent numbers up to 19, counting up to 39.

As they reach the count of 39 again another dot is added to the second position. Any number higher than 19 units in the second position is written using units of the third position.

A unit of the third position is worth 400 (20 x 20), so to write 401 a dot goes in the first position, a zero in the second and a dot in the third. Positions higher than the third are also multiplied by twenties from the previous ones.

Mathematical count							
20	21	41	61	122	400	401	8000

Examples of Maya horizontal positional value representations.

The Maya only made one exception to this mathematical order of tiers and that was with their calender calculations. For example, the Haab' calendar's third position only has a value of 360 instead of 400. This is because the calendar only calculates 18 values of 20, or more accurately, the eighteen 20-day uinals (months) of the Haab' year of 360 days (K'in).

Calendric count

| 20 | 21 | 41 | 61 | 122 | 360 | 361 | 7200 |

Examples of Maya horizontal calendric value representations.

The Maya names for their numbers are as following:

0 - xix im	10 - lahun		
1 – hun	11 - buluc	20 - hun kal	400 - hun bak
2 - caa	12 – lahca	40 - ca kal	800 - ca bak
3 - ox	13 - oxlahun	60 - ox kal	1200 - ox bak
4 - can	14 - canlahun	80 - can kal	1600 - can bak
5 - hoo	15 - hoolahun	100 - hoo kal	2000 - hoo bak
6 - uac	16 - uaclahun	120 - uac kal	8,000 - pic
7 - uuc	17 – uuclahun	140 - uuc kal	160,000 - calab
8 – uaxac	18 – uaxaclahun	200 - ka hoo kal	3,200,000 - kinchil
9 – bolon	19 – bolonlahun	300 - ox hoo kal	64,000,000 – alau

25

Each set counted by twenties.

21 = hun-tukal = 1 + 20
22 = ca-tukal = 2 + 20
23 = ox-tukal = 3 + 20
24 = can-tukal = 4 + 20
25 = ho-cakal = 5 to 2 x 20
26 = ua-ctukal = 6 + 20
27 = uuc-tukal = 7 + 20
28 = uaxac-tukal = 8 + 20
29 = bolon-tukal = 9 + 20
30 = lahun-cakal = 10 to 2 x 20
31 = buluc-tukal = 11 + 20
32 = lah-ca-tukal = 12 + 20
33 = ox-lahun-tukal = 13 + 20
34 = can-lahun-tukal = 14 + 20
35 = ho-lahun-cakal = 15 to 2x20
36 = uac-lahun-tukal = 16 + 20
37 = uuc-lahun-tukal = 17 + 20
38 = uaxac-lahun-tukal = 18 + 20
39 = bolon-lahun-tukal = 19 + 20

Numbers held great significance in the Maya culture. For example, the number 20 signifies the total number of digits a person has: 10 fingers and 10 toes, or five digits on four limbs. As all five digits on a single limb is $1/4^{th}$ the value of a whole of 20, or the value of tier 1.

The number 13 refers to the number of major joints in the human body where the Maya believed disease and illness entered the body. These joint locations were: one neck, two shoulders, two elbows, two wrists, two hips, two knees and two ankles for a total of thirteen.[11]

It's these two numbers, 20 and 13, that are used to make up the Tzolk'in calendar. The Tzolk'in is believed to be the first calendar used by the Mayas. The number 13 is also the number of levels in heaven where the Maya believed the Sacred Lords ruled the Earth.

Chapter 3

The Value of Zero

The ancient Maya had discovered and used zero. They usually represented the value of zero or null with the symbol of an ovular shell. The Long Count calendar requires the use of a zero as a place holder within its vigesimal numerical system. There have been many different glyphs that were used as a zero symbol by different scribes for marking Long Count dates.

Glyph writing was a respected form of art to the Maya. At Chiapa de Corzo, Mexico, the earliest known use of glyphs being used as zero was discovered on 'Stela 2' located there which dates to 36 BC.

The concept of zero is attributed to first being understood and utilized by the Hindus. The Hindus were also the first to use the concept of zero in the way it is used today. A symbol was required in positional numbers to mark the place of a power at the base of a value that was not actually occurring. To mark no value in a value position. This was indicated by the Hindu by a small circle called a "Shunya." This is the Sanskrit word for 'vacant.'

By the middle of the second millennium BC, in the ancient civilization of Babylon the lack of a positional value for zero was indicated by a space between their symbols of numerals.

The Babylonians used a sexagesimal counting system that had the value of 60 at its base. This was the same sexagesimal system that was also used by the ancient Sumerians during the third millennium BC. They had passed down their system of mathematics to the ancient Babylonians.

In 498 AD, the Indian (Hindu) mathematician and astronomer Aryabhatta introduced the decimal system when he stated, "Sthanam sthanam dasha gunam." This statement means, "place to place in ten times in value." This may have been the origin of the modern ten-based decimal value system used today.

The ten number based system used with the Hindu decimal zero was adopted by Arabian mathematicians. They had further modified it and introduced the decimal system and the concept of zero to the Europeans during the Middle Ages.

There are two concepts of zero. One concept is that as being a placeholder in the numbering system to indicate the absence of numbers in a numbering column. This usage was known by the ancient Babylonians and surprisingly, also by the Maya some centuries before. The zero representation used by the Maya civilization didn't look like ours and was used slightly differently because their number systems stacked and wasn't ten-based system.

The other concept of zero is that as being a "null number', or what you get when you subtract 1 from 1. Instead of being a placeholder for the absence of a value, it is the value of nothing. This concept was not developed until some time later. It is estimated to have been realized by at least after 600 AD, but nobody knows exactly who had came up with this concept, or exactly when.

It is speculated that it could have been the Arabic mathematicians, but there is no documentation to be that certain.[12]

Since the eight earliest Long Count dated artifacts appear outside of the main Maya homeland, it is assumed that the use of zero in the Americas pre-dates the Maya civilization and was possibly the invention of the Olmecs. Many of the earliest Long Count dates were found located within the center areas of the Olmec civilization which had already ended by the 4th century BC. These Olmec artifact dates are several centuries before the earliest known Long Count dated artifact that has yet been found.

In addition to understanding the concept of 'zero,' there are some examples in the Mayan language that tell us that the Maya also understood the notion of infinity.

Here are some examples:

- *"Hun tso'dz'ceh,"* to count the hairs a deer has.
- *"Maxocbin,"* infinite in number.
- *"Hunhablat,"* countless.
- *"Picdzaac(ab),"* long number, countless.
- *"Ox'lahun D'zakab,"* eternal thing.
- *"Hunac,"* countless times.

The Maya used a vigesimal numerical system that's based on sets of 20. In a true twenty based system, the first number denotes the number of units up to the value of 19. The next set would denote the number of 20's up to 19 times until the sum value is 400. The next set of numbers are the 400's up to 19 times and so forth up to the next set.

This rule of the vigesimal is followed by the

Maya with the exception of when it was used for calenders in the third place value only the numbered of to the 360's, instead of the number of the 400's. This is because of the 18 20-day uinals that make the 360 days in the Haab' year.

This vigesimal mathematical system is used in the writing used by the scribes that wrote the Dresden Codex. It's the only math system of the ancient Maya for which we have any written evidence of. This is the number system used by the Maya priests and astronomers for celestial and calendaric calculations.[13]

Besides calendars and dating, the Maya needed a counting system they could easily use on a day to day basis. A counting system that would have been used by merchants and traders. This had to be a commonly known numbering system that was used in daily speech when communicating amounts.

The Maya commonly used a dot to represent one value using a cocoa bean or a pebble for counting. We can speculate that they may have perhaps used a stick as a horizontal bar to represent 5 and other special symbols to represent the values of 20, 400, or 8000, an amount of which we know they called a 'pic'.

Although no trace of such a counting method remains, we can reasonably speculate that the Maya used a simple numbering type counting system of such as pebbles and sticks. The count could be higher with this method with higher numbers being calculated by repeating or removing the sticks and pebbles as many times as was needed to make the count.[13]

The Maya vigesimal 20 based counting system has been found in use through numerous different archaeological discoveries. The Maya used mathematics for a wide spectrum of things. However, it should be noted that it is extrapolated by some that the Maya did not have methods of multiplication for their numbers and definitely did not use division of numbers. This cannot be true as the Maya counting system is certainly capable of being used for the operations of multiplication and division.[13]

The Maya vigesimal system still tends to confuse people. Counting that goes 1, 20, 400, 8000, 160000, etc., can seem complicated and confusing when you're trying to figure out how it was useful to anyone except to very 'bright' Maya?

As mentioned earlier, the decimal system is based on ten which we can get by counting our fingers, 1 - 10. Whereas, the Maya counted all the way to twenty by counting all their fingers AND toes to 20. We use the decimal system and count in sets of 10, the Maya used the vigesimal system and counted sets of 20. Each set of 20, goes up the next level and is then counted in sets of 20 again and so forth up each tier or level.

That can get rather confusing when you're not used to counting like that. What about a way to count simple things in "Maya 20 count way" without being very good at it?

Chapter 4

The Four Slave Example

In the "Four Slave" example, we make the assumption that the ancient-era individual counting is using the fingers and toes of four slaves they have in possession to count out a 'pic' of cacao beans. A "pic" in Mayan is 8000. The ancient Maya packed cacao beans in sacks of 8000, which is a pic. Thus, we assume it takes four Maya slaves to count and pack a 'pic' of cacao beans, without actually knowing how to count. Here is how we do it:

To count 8000 cacao beans with 4 slaves that cannot count, all you need to do is make sure they have all their fingers and toes.

- Take "slave 1" and have them pick a cacao bean for each finger and toe they have (20). When they have a cacao bean for each finger and toe, they put that sum into a single pile of cacao beans. They then pass their pile of cacao beans to the next slave, "slave 2."

- "Slave 2" then keeps a stack of cacao beans they get from "Slave 1" for each finger and toe that they have. Once they have a stack of cacao beans from "Slave 1" for each finger and toe they have, they combine it into one stack. They then pass their stacks to the next slave, "Slave 3," whom cannot count either but also has all their fingers and toes.

- "Slave 3's" job is to watch what the other slaves are doing and when "Slave 2" has enough stacks for each finger and toe they have, "Slave 3" gets passed "Slave 2's" stack. "Slave 3," whom proudly caught on easily, then makes a stack for each finger and toe they have from the stacks "Slave 2" passes them.

- "Slave 4" has the easiest job, all he has to do is wait until "Slave 3" has enough stacks from "Slave 2" for each finger and toe. Once "Slave 4" gets passed "Slave 3's" pile, "Slave 4" only has to put the stack of cocoa beans he gets from "Slave 3" and put them in a sack and let his master know, "they have a 'pic' of cacao beans."

None of the of them may be able to count past 20,

much less to 8000, but they can still accurately pack 8000 cacao beans - as long as all the slaves have all their fingers and toes.

When you look at it in this perspective, Maya arithmetic isn't so difficult and you can see how it can be simplified and used in everyday trade. In addition to the significance that they understood the concept of 'zero' before Europeans did, which opened the door for advanced mathematics and science.

Personally, I was pretty satisfied with the "Four Slave" example. It made sense in a primitive society, making that ever regretful assumption that they were all thinking in very basic forms. We tend to underestimate human curiosity and when we do that, we' make false assumptions and forget that the Maya had complex planetary calculations that we've just been able to calculate and verify with the assistance of modern computers. The Maya didn't have computers, powerful telescopes, or any the modern equipment and teams of scientists that we have in the modern age. Or did they?

Well, as wonderful a fantasy as it may be, I am pretty certain aliens from another world or dimension didn't visit and teach the Maya the mathematics necessary to calculate the universe's celestial bodies. It is also safe to assume, aliens probably did not arrive in flying saucers and build the temples and structures of the ancient Maya civilization.

The Maya certainly needed a higher understanding of mathematics beyond basic cacao bean counting to accurately calculate the paths of celestial bodies, timing and patterns. You need higher math for the engineering necessary to build the structures that they built, many that stand to this day. After all, you can only go so far with lucky block stacking.

With this in mind, we need to consider a simpler system to count and perform basic arithmetic. Just in case we don't have any slaves or our slaves are missing some fingers and toes.

Chapter 5

The Grid System

If you don't have any slaves left over from the last raid on the neighboring city, you could always use the grid system to count and do your arithmetic. The "Grid System" is basically what it sounds like. You use a system of grids to do your addition and subtraction.

The grid system works extremely well in the Maya stacked vigesimal system. This is also a wonderful exercise tool that can be used to teach Maya numbers, adding and subtraction, and help to better to better understand the Maya counting system in general.

You start by making a simple grid. You can make a grid the ground using sidewalk chalk or by using masking tape on a table. Gather up some stones and small sticks. Or you can use items that you may have around, such as beans and Popsicle sticks, for example. Begin by making two grid squares, side by side. Start the count by placing one bean in each grid square.

Examples how the Grid System works in simple Addition:

A single bean or 'dot' in each square is how the value of "1" is measured and conveniently, also written in Maya. The value of one is a single dot.

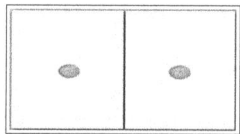

For simple addition, we have a single bean or dot in each square representing 1 + 1.

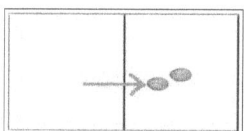

We add our 1 + 1 by sliding the beans from the left grid (grid A) into the grid on the right (grid B), making our total count equal to 2 (1 + 1 + 2).

We'll try another example by adding 3 + 3. We take three beans, representing the Maya symbol for the count of 3 in each of the two grid squares.

44

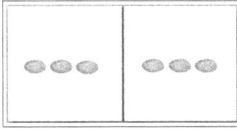

Here, we have 3 in grid A and 3 in grid B to add.

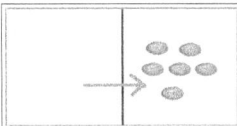

We take all the beans in grid A and slide them all into grid B to be summed up.

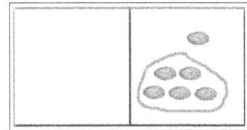

Remember, in Mayan, five dots equals a bar. So we gather 5 beans and we exchange them for a bar, or stick.

After removing the 5 beans and exchanging them for a stick (bar), we have the represented sum of 6. This is also properly represented in Maya writing, as six is written with one dot over one bar.

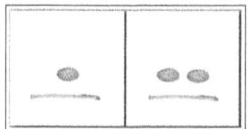

In this example, we will add 6 and 7, represented by 1 bar and 1 dot in grid A and 1 bar and 2 dots in grid B.

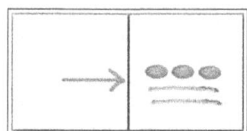

We again, push all the sticks and beans from grid A into grid B. We then add the number of bars (sticks) and dots (beans) to get a total count of 13 (two 5's (bars) + three 1's (dots) = 13).

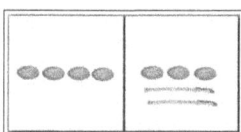

Here, we shall add 4 and 13.

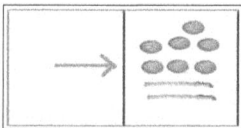

We move all the dots and bars from grid A into grid B.

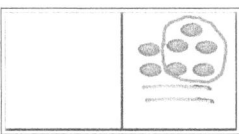

Remember, gather every group of 5 dots and exchange them for 1 bar.

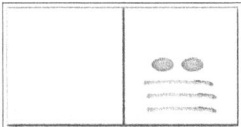

We add the exchanged bar and have our sum of 17.

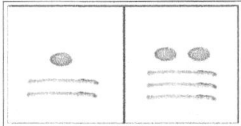

Let us continue by trying to add 11 and 17 together.

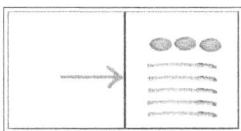

Move all the dots and bars into grid B.

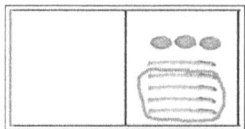

We now remove 4 bars to make 1 dot for the next tier, because four 5's equals 20 and the count moves up a value in tier per 20 values on the previous tier.

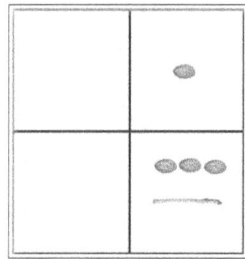

We remove the 4 bars from tier one and exchange it for a single dot in tier two representing the value of 20. We add a single dot in the 20 tier plus 3 dots and a bar in the ones tier and our sum is 28 in total (20 + 5 + 3 = 28).

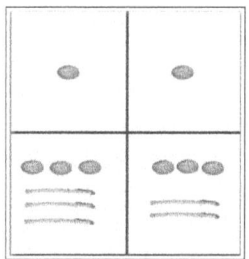

Here we begin our addition using two tiers (the 1's tier and the 20's tier). On tier two, in both grids, we have a dot in place of the 20. On tier one, we have our sums 18 and 13, making the total as 38 + 33.

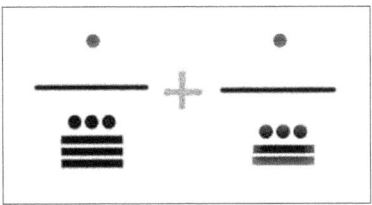

Example of 38 + 33 written in Maya.

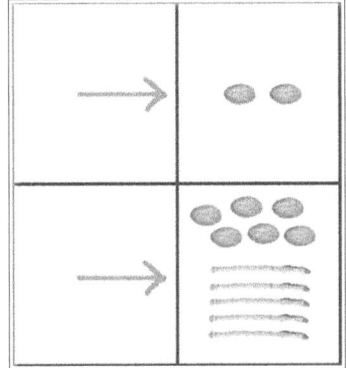

Again, we move all the beans and bars from grid A into grid B, staying in the tier levels of each grid.

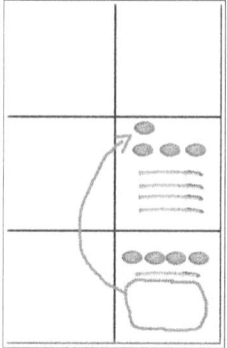

Starting on the bottom tier, we remove each group of 4 bars and exchange them for a bean to place in the second tier group.

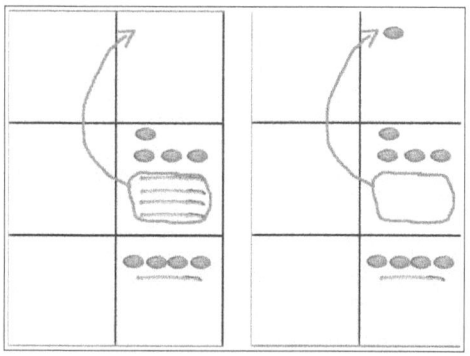

Next, on tier two, we again remove any groups of 4 bars and exchange them for a dot for the next level tier up, tier 3 (20^2).

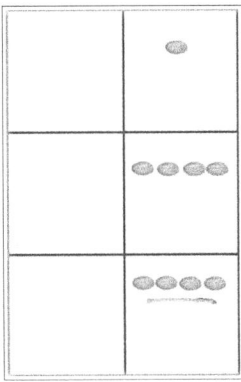

And now we have, as it would also look if written in proper Maya, the sum of 489.

$$(400 + 1) + (20 + 4) + (4 + 5) = 489.$$

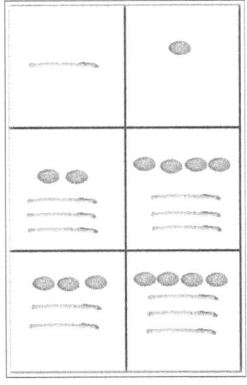

Let us try another on three tiers.

We shall add 2373 + 799, without counting past 5 to get our total.

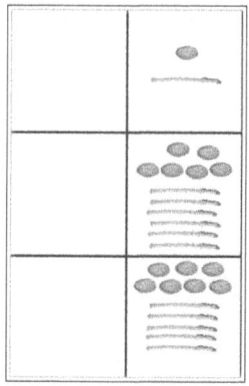

As before, we move all the dots and bars from grid A and slide them all into grid B, staying in whichever tier level they were in.

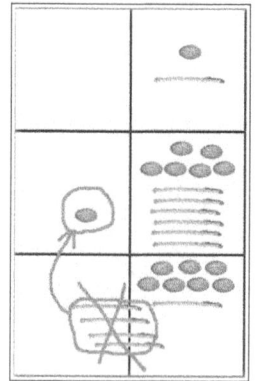

We begin by starting on the bottom tier and add groups of 4 bars and exchange them for a dot to the next tier level up.

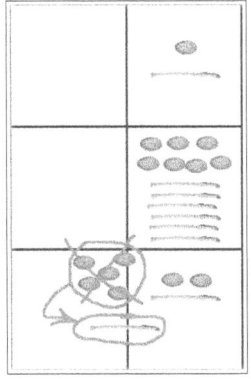

We also exchange all the groups of 5 dots for bars, if you hadn't already done so.

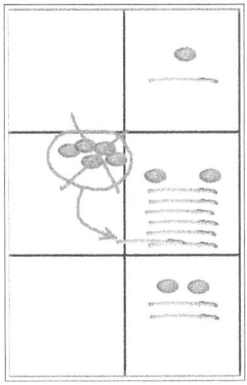

We continue by going up the next level tier and add a bar for every 5 dots.

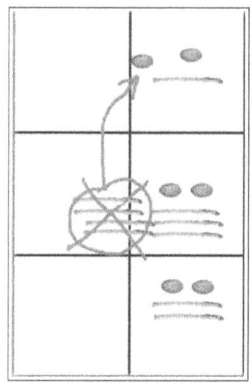

We also add a dot to the next level tier for every 4 bars that we have.

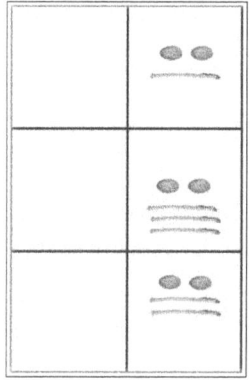

And now we have the sum of 3152.

Chapter 6

Subtraction

The grid system works with subtraction as well as it works with addition. The except is when we subtract in Maya, we borrow from 20, instead of 10 like in the decimal system.

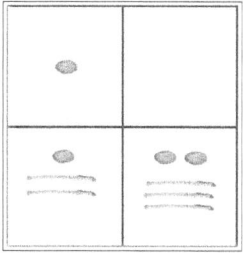

In this example, we shall subtract 17 from 31 (31 – 17).

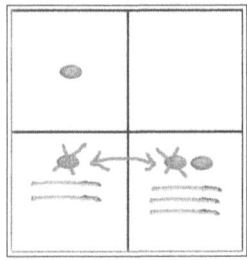

Subtract first the lowest value, in this case the ones. Remove one dot in grid B for one dot in grid A. To Subtract, you remove a value from one grid and match and remove the same value in the other grid.

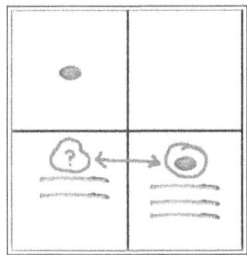

At this point, we've run out of dots from grid A. To resolve this, we remove 1 bar in exchange for 5 dots in grid A.

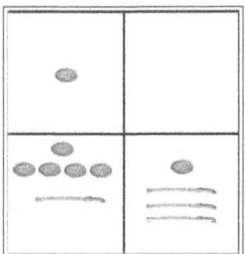

Here, we're exchanged 1 bar in grid A for 5 dots.

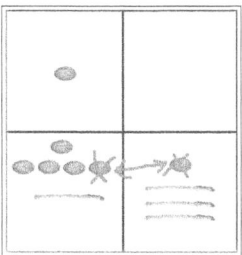

We then continue removing dots in grid A for every dot we have left in grid B.

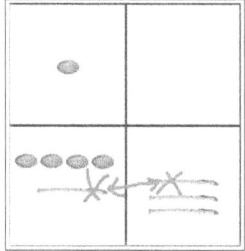

And continue subtracting by removing every bar from grid A for every bar in grid B.

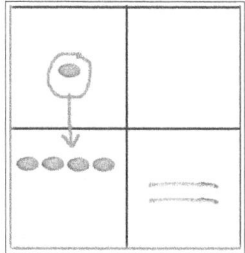

Now that we've removed every bar in grid A for every bar in grid B, we find that we are short bars in grid A because we still have 2 bars in grid B to subtract. In this case, we borrow from tier two and subtract 1 dot (value = 20^1) and add 4 bars to tier one in it's place.

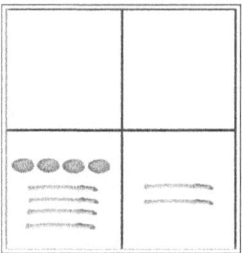

By borrowing from tier two, we now have bars in grid A to remove for the remaining 2 bars we have left in grid B.

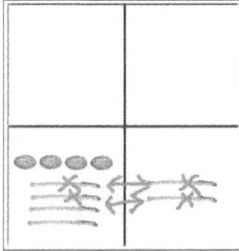

Continue subtracting bars in grid A for every bar grid B has.

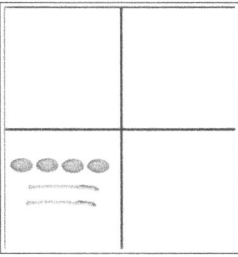

Now that we've removed every dot and bar from grid A that we had in grid B, we finish with our total of 14.

$$31 - 17 = 14$$

Chapter 7

The Finger Method

Simple calculations can also be made using the Maya vertical vigesimal system without grids, beans, or bars. If we were conducting trade or simply needed to do some simple and quick math. Gathering slaves to help count using their fingers and toes is not very convenient and what if we were a lower status merchant or priest and didn't have any slaves to help us count.

Making a quick "grid system" on the ground would work, but we'd still have to gather sticks and stones to do the counting and that's no good in a hurry.

So we must reason that the Maya would have simplified their system of basic counting so they could count on their fingers, like we do with our decimal system of 10s. Simply using all our fingers and toes to count to 20 is NOT convenient, especially if you had your shoes on. The Maya simply had to have an easier way to count through their base system of 20 with just their hands.

The Maya didn't count linear in 20's like our current modern system does with 10's. Additionally, the Maya didn't have specific characters to designate specific numbers, like the Arabic system that uses specific symbols for 0, 1, 2, 3, 4, 5, 6, 7, 8, 9. They used a system of dashes (lines) and dots to represent a value. A tiered stack of lines and dots to represent the sum of a number, not actually having a specific symbol to represent the number.

In fact, the Maya method of writing numbers never counts past four dots or three bars. The symbols always changes and goes up to the next value. A count past 4 dots is exchanged for a bar (value=5) and 4 bars (4x5) is a dot on the next tier valued at 20. This is how they could have huge, long count numbers with relatively few symbols to represent the amount. The Maya didn't count using all ten fingers to reach the count of 10, they only used two fingers to represent 10.

Assuming the Maya counted in the same manner as we do, counting each finger with a value of 1. They would stop with using four fingers and then would use a seperate finger to represent the value of 5. WJereas, when we use our fingers to count, we use all five fingers on our hand to display the vlaue of 5. The Maya only needed to use their fingers to count up to four dots, then count a bar and then bars with dots and so forth. The value of 6 was one bar and one dot, or one finger from each hand, representing a bar and a dot.

To count using the Maya vigesimal system, on one hand each of your four fingers represents a value of 1. On the other hand, each one of your fingers represents a value of 5. When you count, you count on your hand with the value of 1 in sequence of 1, 2, 3, 4. then drop those and add a value of 1 on the other hand.

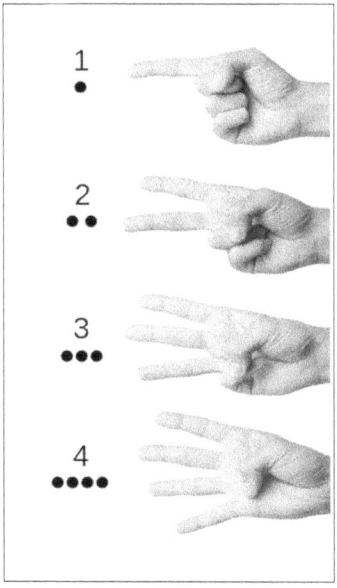

On the first hand, count: 1, 2, 3, 4.

When you reach 5, close the fingers on the first hand and you raise 1 finger on the other in the value of 5.

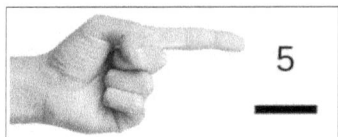

One finger on the other hand represents five fingers on the first hand.

On the first hand, you continue the count: 6, 7, 8, 9.

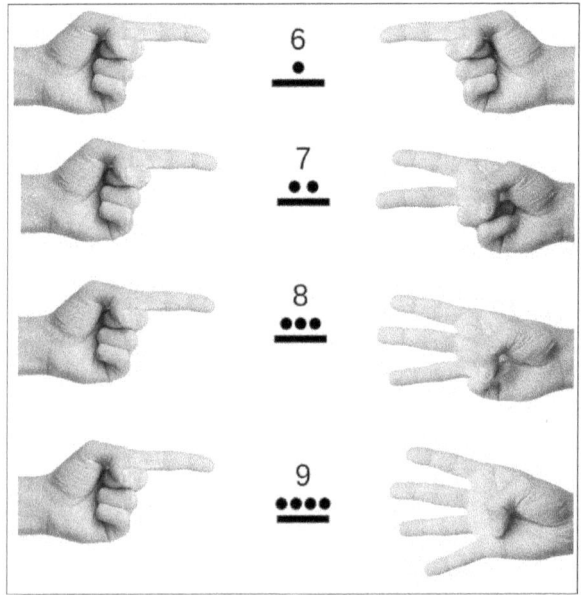

Each finger on the left hand = 5, and each finger on the right hand = 1.

When you reach 10, all the fingers on the first hand go down and the second hand now displays two fingers representing the value of 10 (5 + 5).

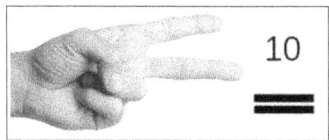

The value of 10 is represented by two fingers on the second hand (5 + 5 = 10).

On the first hand, with two fingers on the second hand out, you continue the count: 11,12,13,14.

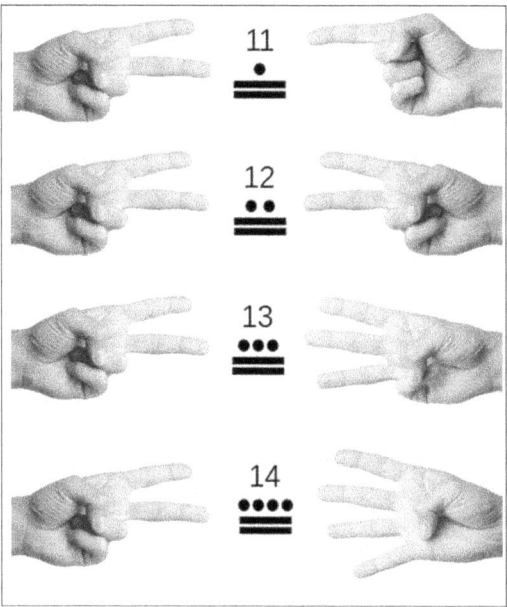

Then at the count of 15, raise an additional finger, totaling to three fingers which represents the value of 15 (5 + 5 + 5).

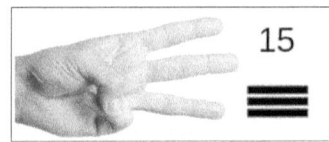

Thus, with three fingers out on the second hand, we continue the count on the first hand with: 16,17,18,19.

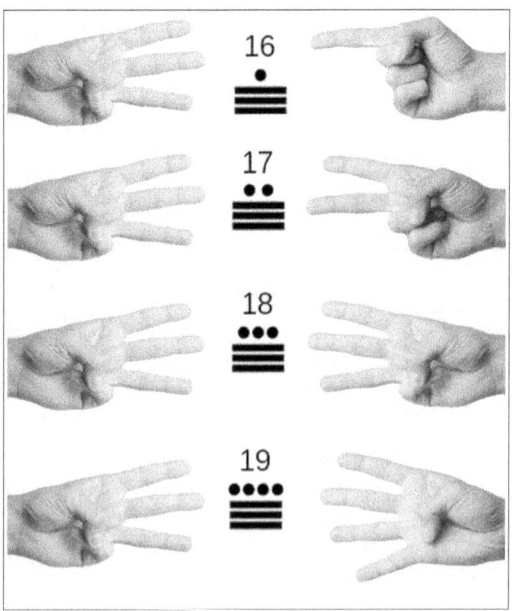

At the count of 20, you close the fingers on the first hand and display all four fingers (or alternatively use your thumb) on the second hand to the value of 20 (5 + 5 + 5 + 5 (or with the thumb representing 1 value on the second tier, which equals 20)).

Wow, that is amazing, I think we just unlocked the mystery to a Maya simplified counting system. We can easily count to 20 with just our two hands and use the same dot and dash system to represent counts of 1 for dots and dashes as 5s.

The count is not in linear values counting to 20, it is in the value of ones and fives and their relative position of counting up in segments of 4 and 5, with the sum of 4 or 5 being the next value, represented by the next symbol and position. Simply, when four of the the "fives" are used up in their sequence, they go to the next level or tier.

The value of one and five also changes as it goes up each level. This is why there are no number symbols in Maya counting, only dots for values of 1 and bars that represent the value of 5 dots. There is no "20," there is only one dot on a tier, which merely represents the value of four bars on the previous, lower tier.

Using the 'finger method,' you can even alter the system slightly and give each thumb the value of 20 (one dot each on tier two in Maya thinking). With both thumbs having the value of 20 each, equaling 40 (2 x 20). The value of 'one' for each finger on the first hand (4 x 1), and the four fingers on the other hand with each valued at 5 (4 x 5). We can easily count to 64 (20 x 2 + 4 x 1 + 4 x 5) with just two hands. Whereas, previously we could only count to 10 using the same fingers with the decimal system.

Interesting as well, that we can count to 64 with our hands using the Maya system. It appears to be a number significant to a computer's binary method of counting using on and off switches that we call bytes. There are 8 bits to a byte. A row of 8 bytes is 64 bits. Of course this multiplies up to 128, 256, 512, etc., which I am sure you recognize as numbers used in computers as: gigabytes (GB), megabytes (MB) and Kilobytes (KB).

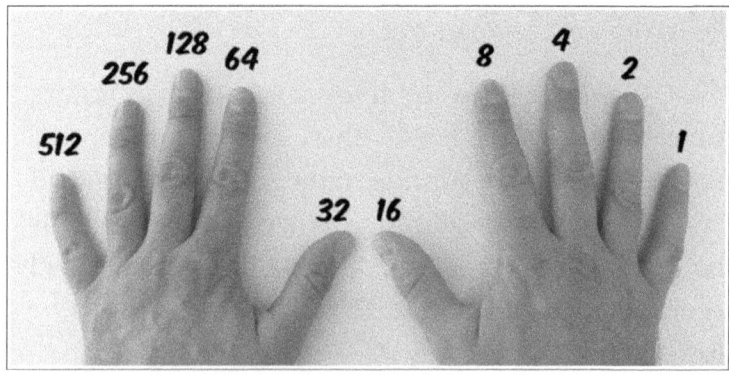

Number values for using fingers to count in Binary System.

A computer or calculator counts using a binary system of off and on switches, each with a value of 0 or 1. This is broken down into a sequence of 1, 2, 4, 8, 16, 32, 64, 128 , 256, 512 typically.

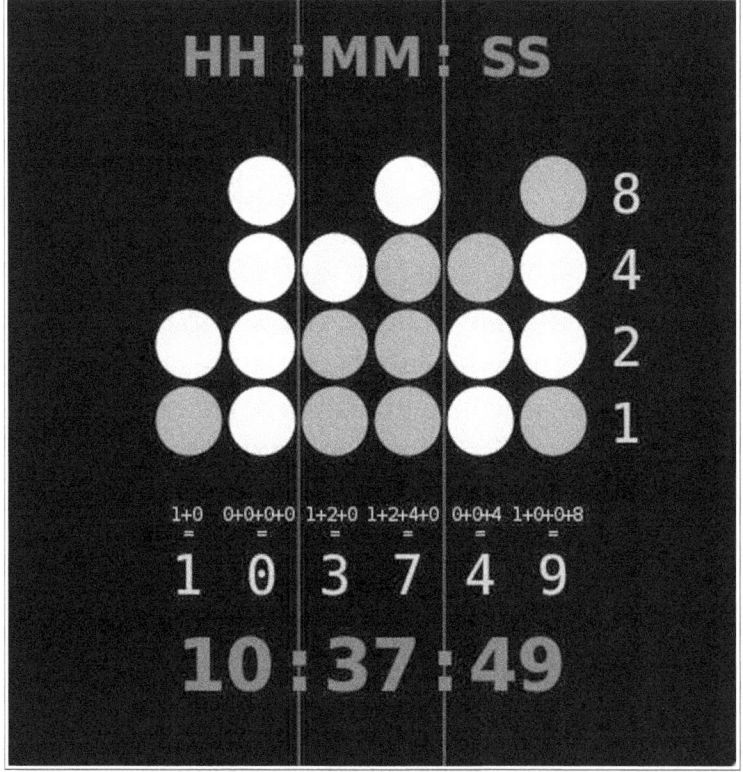

Visual explanation of a binary clock..[14]

A binary clock might use LEDs to express binary values. In this clock, each column of LEDs shows a binary-coded decimal numeral of the traditional sexagesimal time.

Without going into greater detail about binary counting and how to calculate using the binary system, we can see how the Maya counting system is similar to the binary system computers use. This means the Maya could use their system to perform complex calculations needed to accurately plot astronomy, calendars, and the mathematics necessary for the engineering feats they had achieved.

The Maya system of counting with your fingers is great, but you do eventually run out of fingers and toes and would like an improved way to do basic arithmetic. This where we take it to the next level and build an abacus (plural abaci or abacuses). An abacus is also called a 'counting frame' and is a calculating tool that is used for performing arithmetic processes.

Chapter 8

The Maya Abacus

The Mesoamerican abacus is called a "Nepohualtzintzin."[15] The arrangement of the Maya Abacus, or Nepohualtzintzin, is with seven beads or balls (or cacao beans) per level. Every level higher unit sum is equal to the sum of all the units of one level less than it. the first level being twenty units of one. alue the sum of level up in the Maya counting system, represented in illustration 1 by white lines, labeled: A, B, C, D, E, and F.

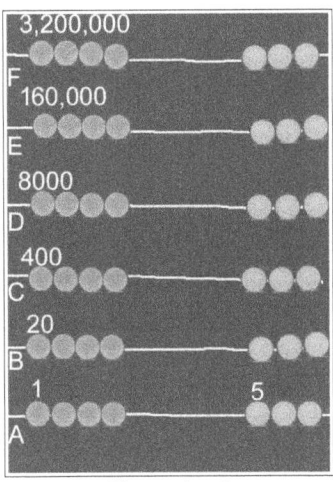

illustration 1

75

The represented value of the beads on each level in these examples are:

Blue beads = 1 unit.

Red beads = 5 units.

Each level's unit increases in value each level up, as represented by white lines in *illustration 1*, as follows:

Line A = 1,

Line B = 20,

Line C = 400,

D = 8000, E = 160000, etc.. as shown in *illustration 1*.

Using the Nepohualtzintzin is much like using any other the World's various abacuses through time. You begin the count on the first line, line A (level or tier 1), by sliding a blue bead to the middle of line, counting to 1 (see *illustration 2*).

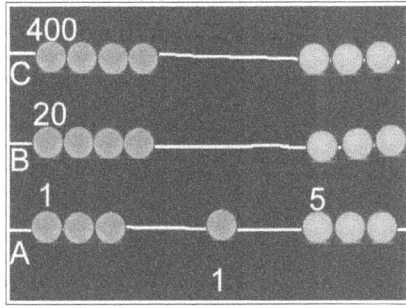
illustration 2

The absence of any beads in the middle of any line would be zero and represented with a shell in written Mayan. Continue counting up with your Maya Abacus by adding blue beads.

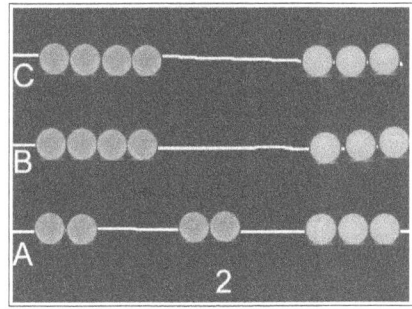
Illustration 3

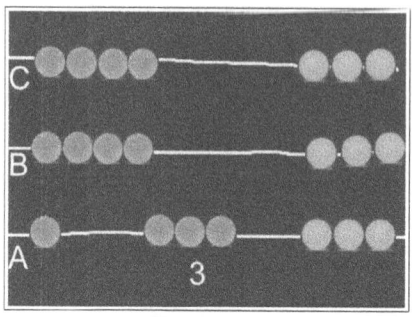
Illustration 4

When you've used all four blue beads counting to 4,

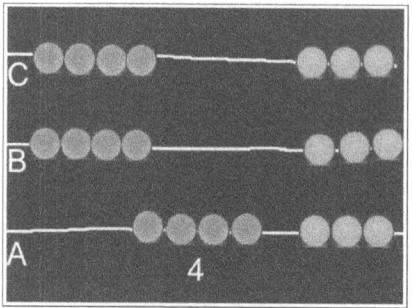
Illustration 5

Slide all the blue beads back to their original position and slide one red bead, value 5, to the middle to make the count at 5. as displayed in *illustration 6*.

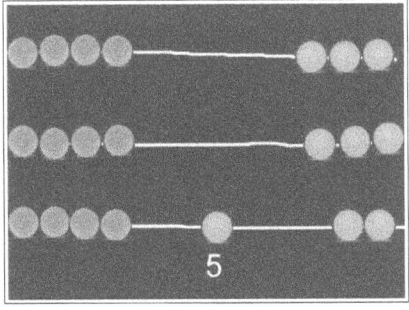

Illustration 6

The count continues by adding a blue bead next to the red bead, making the sum 6.

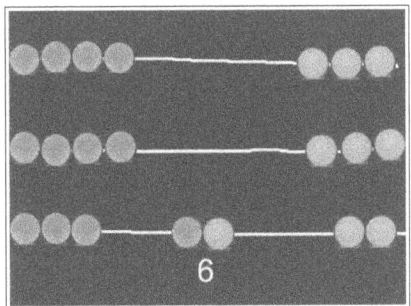

Illustration 7

Continue adding on line A with blue beads, counting up 7, 8, and 9. Ten is then counted by returning the blue beads to their original position and sliding an additional red bead (value is 1 red unit = 5 blue units) to the middle (5 + 5 = 10).

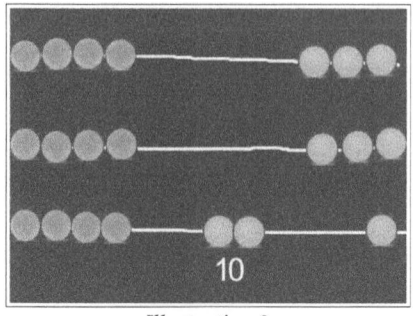

Illustration 8

An additional red, adds 5 to the count, making the sum 15.

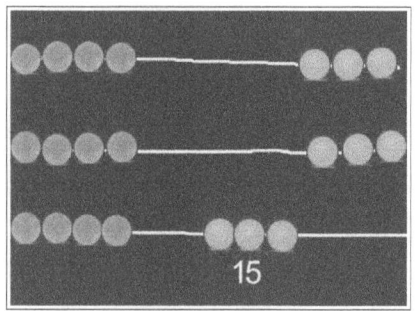

Illustration 9

The count continues by adding blue beads, counting up 16, 17, 18, and then with all red and blue balls in center equals 19. (*illustration 10*).

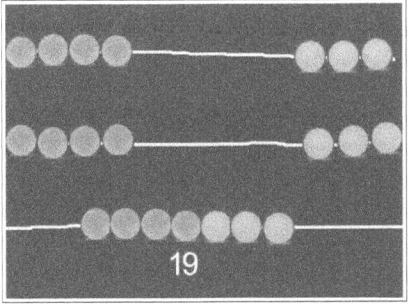

Illustration 10.

Continue counting up to 20 by sliding all the beads on line A back to their original start positions and slide 1 blue bead on line B (valued at 20 per unit) to the middle.

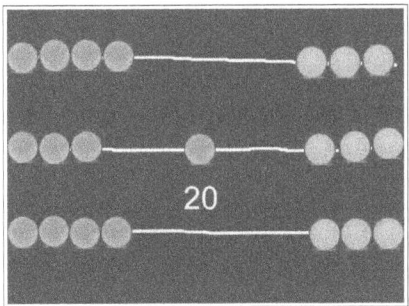

Illustration 11

$$\frac{\bullet}{\text{🐚}} = 20$$

A single blue bead in the middle of the second level (line B) brings the count to 20. When writing this number, you would add a shell on the bottom tier to represent zero as a place holder.

Increase the count by adding 1 blue bead on the first line, line A (illustration 12).

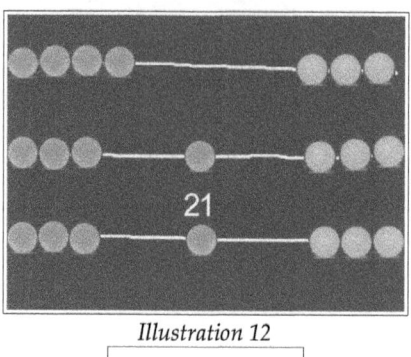

Illustration 12

●
   = 21
●

Some additional examples of using the Maya Abacus count:

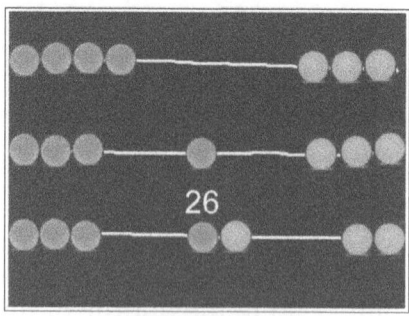

Illustration 13

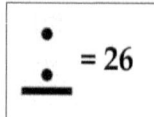 = 26

Remember, each blue bead is the value of 1 unit and each red bead is the value of 5 single units, with the value of each unit starting at 1 on the first level and increased 20 times the value of a single unit value of the line (1, 20, 400, 8000, etc.).

In *illustration 14*, one red bead (value = 5) on line two, representing 5 blue beads (value 1 unit = 20), shows the total of 100.

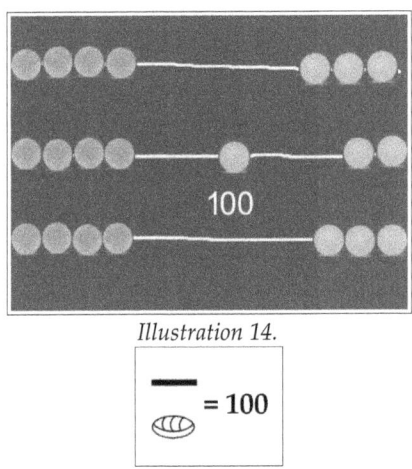

Illustration 14.

We then increase the count to 111 by adding two red beads on line A, valued at 5 units each and 1 blue bead on Line A, valued at 1 unit (100 + 11 = 111).

83

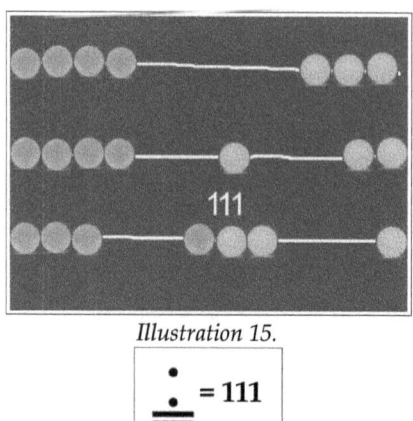

Illustration 15.

We set the count at 400 by adding 1 blue bead on line C (3rd level). We know that the value of the single blue unit bead on the third level are worth 400 because the sum of a units on the 3rd level are equal to the sum of all the units from the previous level, which were valued at 20 per unit, which is the sum of all the units on the 1st line which are valued at 1.

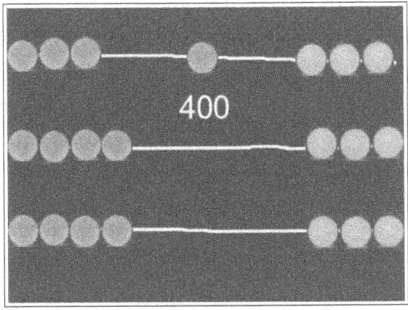

Illustration 16.

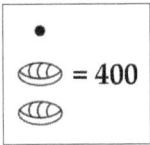

When we write 400 in Mayan, we write a single dot on the 3rd level and a single shell on each level under it, on the 1st and 2nd levels, as seen under *illustration 16* above.

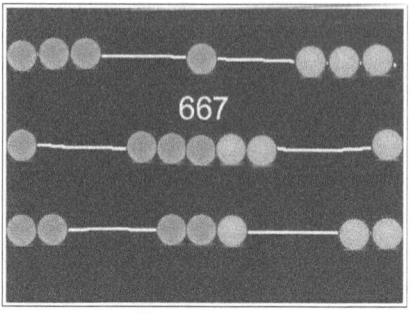

Illustration 17.

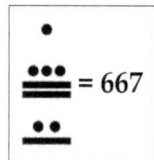

(1 x 400) *Line C*

+

(13 x 20) *Line B*

+

(7 x 1) *Line A*

= 667

The development of the Nepohualtzintzin, or Mesoamerican abacus explains the magnitude of understanding that the Mesoamericans had in mathematics.

Knowledge in these mathematics made it possible that they were able to make such exact calculations of universal cosmogony. The Nepohualtzintzin, which essentially was a pre-Hispanic computer, was not only able to make mathematical calculations, but also astronomical and gestation interpretations. The Nepohualtzintzin as an instrument that is similar to other abacus in different cultures, such as the Japanese soroban.

The abacus helps make it possible to perform not only basic operations such as: addition, subtraction, multiplication and division, but it also can be used for complex operations like roots, powers and integral and differential calculus operations.

The word "Nepohualtzintzin" comes from the Nahuatl language and is formed by the roots; Ne - personal -; pohual or pohualli - the account -; and tzintzin - small similar elements. This roughly translates into: counting with small similar elements by somebody.

The knowledge of the Nepohualtzintzin and its use was passed on to students, whom dedicated their entire lives from childhood to mastering and calculating the events and movements of the skies.

Unfortunately, the Nepohualtzintzin and its teachings were among the victims of the evangelizing paranoia of the Spanish Conquest.

The Nepohualtzintzin proves that Mesoamercian cultures already had great capabilities in scientific and technological developments prior to the arrival of the Europeans.

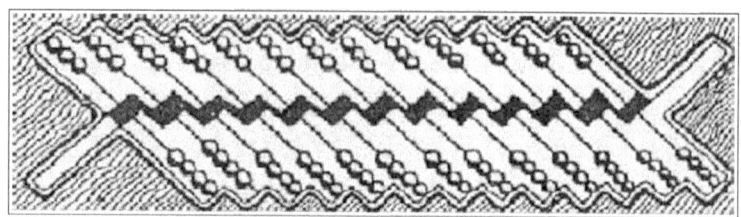

A typical Nepohualtzintzin usually has 13 rows with 7 beads in each row.

A Nepohualtzintzin has 13 rows with 7 beads in each row, which makes for a total 91 beads.

- The 91 beads in the Nepohualtzintzin represent the number of days in a season within the year.

- Two Nepohualtzitzin make a total of 182 beads, which is the number of days of corn's cycle from sowing to harvest.

- Three Nepohualtzintzin making for a total of 273 beads and is the number in days of a human baby's gestation time from conception to birth.

- Four Nepohualtzintzin complete the cycle of a year's time, minus a day and a quarter.

The Nepohualtzintzin accounts for the absolute precision needed for the higher scientific and mathematical levels that the Mesoamericans had developed many years before the arrival of the Spanish conquistadors.

David Esparza Hidalgo's had rediscovered the Nepohualtzintzin upon finding diverse engravings and paintings of the ancient Mesoamerican math instrument in Mexico. Very old Nepohualtzintzin that were discovered could have also been attributed to the Olmec culture. Some of the ancient abacus that have been found were usually in the shape of bracelets, especially those discovered in the Maya area. One of the ancient Nepoualtzitzin abacus appeared on a painted vase in Guatemala known as the "Nejar Vase."[15]

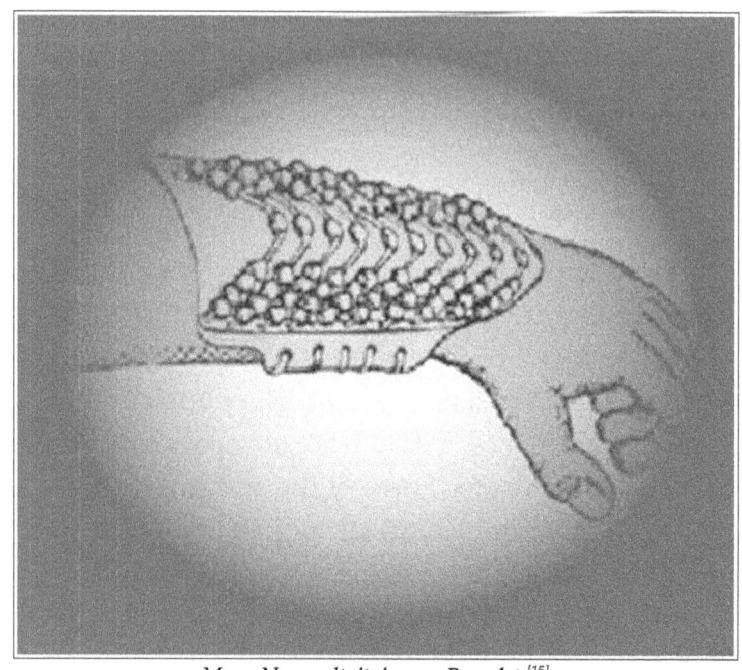
Maya Nepoualtzitzin as a Bracelet.[15]

The existence of the Mesoamerican abacus gives us insight that the ancient Maya already had the sufficient knowledge to devise and handle a device of mathematical complexity and also extend its usage into their daily lives and activities.

Chapter 9

Maya Concept of Fractions

The word fraction comes from "fractus," the Latin word for broken. It represents a part of a whole or any number of equal parts of the whole. Broken or fractured from the whole.

A common, vulgar, or simple fraction consists of an integer numerator, which is displayed above a line (or before a slash), and a non-zero integer denominator, displayed below (or after) that line. The numerator represents a number of equal parts and the denominator indicates how many of those parts make up a whole. For example, in the fraction ¾, the numerator 3 on top, tells us that the fraction represents 3 equal parts, and the denominator, 4 on the bottom, tells us that 4 parts make up a whole.

In contrast to previous claims by researchers, the Maya were familiar with a notion of fractions or "parts of a whole." To indicate parts in general, they used the term "tzuc" which literally means "part." The Mayan words, "tu," "can," "tzucil," and "ban cah," equals the four parts of the World (cah), or the four quarters of the whole World.

For the notion "¼," we find the expressions, "heb" (to open) and "u" (moon).

Some examples:

- "heb u" = moon opening or open moon.
- "hun heb u" = 1/4 moon or moon opening of ¼.
- "ca heb u" = 2/4 moon or moon opening of ½.
- "ox heb u" = 3/4 moon or moon opening of ¾.

For the notion of "½," two possible applications can be found.

First, in distance:

- "Tan coch" = half, in the middle.
- "lub" = "legua" (5.5 km).
- "tan coch lub" = half a "legua."
- "tan coch tu cappel lub" = in the middle of the second legua (5.5 km), or 1 ½ "legua."

Secondly, in time divisions, such as:

- "tan coch kin tu cappel" = in the middle of the second day = 1 ½ days.

"Xel" = dividing the unit in two and subtracting one part. Xel is in fact a negative fraction:

- "xel u ca kin bé" = $-½ + 2$ days = 1 1/2 days;
- "xel u ca cuch" = $-½ + 2$ loads = 1 1/2 loads;
- "xel u cappel lub" = $-½ + 2$ leguas = 1 1/2 legua;
- "xel u yox katun" = $-½ + 3$ katun = 2 1/2 katun;
- "xel u ca kal" = $-½ \times 20 + 2 \times 20 = -10 + 40 = 30$;
- "xel y yox bak" = $-½ \times 400 \text{ (bak)} + 3 \times 400 = 1300$.[143]

From these astronomical and time keeping divisional uses in the Mayan language, the best inference we can gather as the most common use of fractions would be quarters or fifths of a whole that reflects the use of the Maya counting system in it's self.

Typically, a bar being a whole and thus is divided into five equal parts, as five dots equals one bar. So, we're able to assume the Maya also had a regular use of 5ths (1/5, 2/5, 3/5, 4/5, and 5/5 = 1 whole).

However, a better inference would be to conclude that fractional systems were in the Maya vigesimal system, no different that regular counting. That in fact, the entirety of the Maya vigesimal system is already fractional.

There are two kinds of scientists;

1. Those who can extrapolate from incomplete data

So, how is it possible to write and use fractions in the Maya system? On the second tier, the single dot (value = 20) would be the whole number (or denominator). On the first level below it (count 0 - 19), would be the fractions (numerators) of the upper whole.

In this example, is the Maya count for 20 (20¹). We then express this whole number as a fraction using Latin numerals.

$$\frac{\bullet}{\text{🐚}} = \frac{20}{20}$$

We could extrapolate by saying a good way to indicate a Maya fraction would be to use the shell (used for the null or zero placeholder) as the placeholder for an incomplete whole when written as a fraction. We use a shell on top to represent an incomplete whole of the dot, whereas when we use a shell below the dot on the bottom tier to represent a whole count of 20 on the second tier. To represent the fraction, we remove the dot on the second tier and replace it with a shell and the shell is now a placeholder indicating an incomplete whole number.

For example, the fraction 1/20th could be expressed in Maya as:

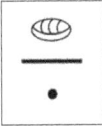

Using the shell (0) as a placeholder for the dot of the whole of 20. The bottom number, the numerator, is the fraction amount of the whole (20¹).

$$\frac{\text{🍞}}{\bullet} = \frac{1}{20}$$

$$\frac{\text{🍞}}{-} = \frac{5}{20} = \frac{1}{4} \text{ (or 0.25 or 25\%)}$$

$$\frac{\text{🍞}}{=} = \frac{10}{20} = \frac{1}{2} \text{ (or 0.5 or 50\%)}$$

$$\frac{\text{🍞}}{\equiv} = \frac{15}{20} = \frac{3}{4} \text{ (or 0.75 or 75\%)}$$

Addition of fractions is carried out the same manner as regular addition by simply adding the sums together.

$$\frac{16}{20} + \frac{12}{20} = 1\frac{8}{20}$$

References

1. "Maya." Dictionary.com Unabridged. Random House, Inc. 19 Jan. 2013.

2. Map of Settlement area of Ancient Maya. Nepenthes, 18 July 2006.

3. Early Civilizations in the Americas: Almanac. (2005). Gale Cengage.

4. Picture of Temple I in Tikal, Guatemala, taken by Bruno Girin. (2005).

5. Sharer, Robert J.; with Loa P. Traxler (2006). The Ancient Maya (6th (fully revised) ed.). Stanford, CA: Stanford University Press. ISBN 0-8047-4817-9. OCLC 57577446.

6. Map of the Mayan Civilization cultural area by © Sémhur / Wikimedea Commons / CC-BY-SA-3.0 2. Martin & Grube 2000, p. 102. Sharer & Traxler 2006, p. 357.

7. The Mayan ruins of Copan Ruinas located near Copan, Honduras. Photograph by Kyle Hammons. July 14, 2009.

8. Lidded effigy container in the form of a diving god ca. A.D. 1500. Late Postclassic Maya. Princeton University Art Museum.

9. Table showing first 20 Maya numbers and their Arabic equivalents. Centro de Estudios del Mundo Maya. Yucatan, Mexico. Maya World Studies Center.

10. Lounsbury, Floyd G. Maya Numeration, Computation, and Calendrical Astronomy. In Dictionary Of Scientific Biography. New York, New York. Charles Scribner's Sons. Volume 15, Supplement 1.

1978. P. 759-818.

11. Hernan Garcia, Antonio Sierra, Gilberto Balam, Jeff Conant, and Hilberto Balam. "Wind in the Blood: Mayan Healing & Chinese Medicine." 1999.

12. McAnany, A. Patricia (1998). "Ancestors and the Classic Maya Built Environme."

13. G Ifrah, A universal history of numbers : From prehistory to the invention of the computer (London, 1998).

14. Visual explanation of a binary clock. Alexander Jones & Eric Pierce. 14 October 2006.

15. David Esparza Hidalgo, Nepohualtzintzin. Computador Prehispanico en Vigencia [The Nepohualtzintzin: a pre-Hispanic computer in use] (Mexico City, Mexico: Editorial Diana, 1977).

16. Kane, Njord. "The Maya: The Story of a People." Spangenhelm Publishing, 2013. 2016.

Cover image: Panel Three at Cancuen, Guatemala, representing king T'ah 'ak' Cha'an. 2005.

Other Books:

The Maya
The Story of a People
by Njord Kane

~

The History of the Maya
by Njord Kane

~

The Vikings
The Story of a People
by Njord Kane

~

The Viking Stone Age
Birth of the Axe Culture
by Njord Kane

~

History of the Norse
by Njord Kane

www.ingramcontent.com/pod-product-compliance
Ingram Content Group UK Ltd.
Pitfield, Milton Keynes, MK11 3LW, UK
UKHW021303180426
11947UKWH00015B/997